FRANCIS FRAMES THE FUTURE

Francis Frames The Future

With the Wisdom of the Past

Story by Matt Bergles, PhD
Illustrated by Rob Peters

MERRY DISSONANCE PRESS   CASTLE ROCK, CO

Francis Frames the Future

Published by Merry Dissonance Press, LLC
Castle Rock, CO

Copyright © 2019 by Matt Bergles, PhD. All rights reserved.
www.mattbergles.com

No part of this publication may be reproduced or transmitted in any form or by any means, electronic, mechanical, including photocopying, or by any information storage and retrieval system without written permission from Merry Dissonance Press, LLC or Sara L. Daigle, except for the inclusion of brief quotations in a review.

All images, logos, quotes, and trademarks included in this book are subject to use according to trademark and copyright laws of the United States of America.

FIRST EDITION 2019

Bergles, PhD, Matt, Author
Francis Frames the Future
Matt Bergles, PhD

Publisher's Cataloging-in-Publication data

Names: Bergles, Matt Paul, author.
Title: Francis frames the future : with the wisdom of the past / Matt Bergles, PhD.
Description: First trade paperback original edition. Castle Rock [Colorado] : Merry Dissonance Press, 2019. Also available as an ebook.
Identifiers: ISBN 978-1-939919-59-5
Subjects: LCSH: Conservation. | Environmentalism. | Francis, Pope, 1936-.
BISAC: JUVENILE NONFICTION / Science & Nature / Environmental Conservation & Protection.
Classification: LCC BV4559 | DDC 241–dc22

ISBN: 978-1-939919-59-5

Book Design and Cover Design © 2019
Illustrated by Rob Peters
Cover Design by Rob Peters
Book Design by Rob Peters
Editing by Donna Mazzitelli, The Word Heartiste

All Rights Reserved by Matt Bergles, PhD
and Merry Dissonance Press, LLC

Printed in the United States of America

SPECIAL NOTE: Appropriate officials within the Archdiocese of Denver have determined this publication does not need an imprimatur.

This book is dedicated to my lifelong friend Jim Brunjak. His extreme generosity made this book, and so much more, possible. Thanks for everything, Jim! Your memory and good deeds will live on through your children and grandchildren.

Hi! I'm Pia. And I'm Pio. We are European turtle doves who live in Italy. We want to tell you a story about a man named *Francis* who is trying to help our planet and all the animals that live on Earth, especially wild animals like us.

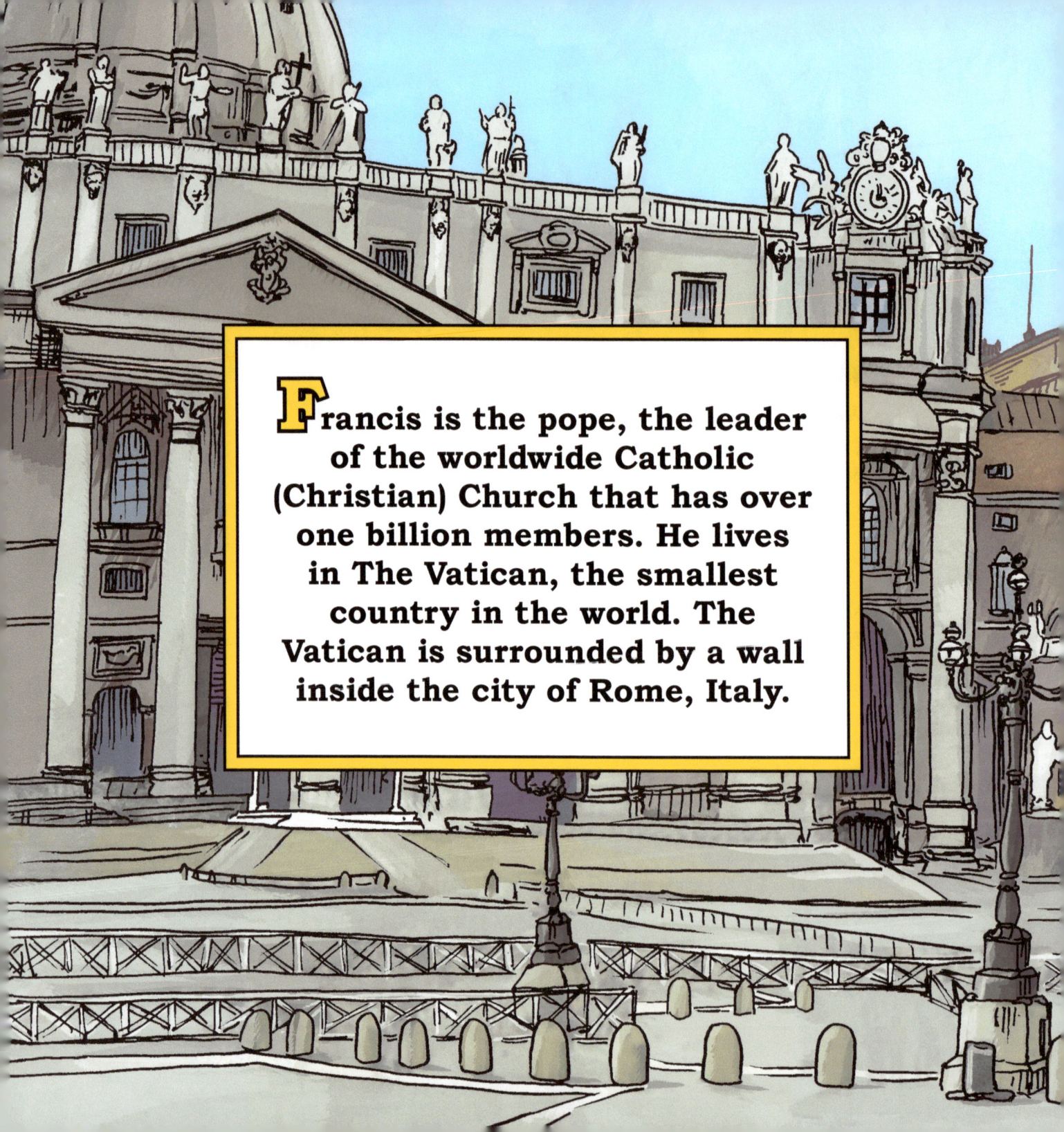

Francis is the pope, the leader of the worldwide Catholic (Christian) Church that has over one billion members. He lives in The Vatican, the smallest country in the world. The Vatican is surrounded by a wall inside the city of Rome, Italy.

Popes take the names of saints. Pope Francis took his name from a very special saint who lived about 800 years ago in Assisi, a small town in Italy. One reason the pope took Francis's name was because of the saint's love for all of God's creatures. Like St. Francis, Pope Francis loves animals and believes people need to do a better job of taking care of these creatures.

There are lots of stories about St. Francis talking to animals. One story describes how Francis told a wolf to stay out of the village of Gobbio because he was scaring the people. From that point on, the wolf and the people lived in peace.

Most of the stories tell about St. Francis's special relationship with birds. They were his favorite animals, and he loved communicating with them.

St. Francis believed that all of nature, from the smallest creatures and plants to human beings, are connected to each other. So today, many people of all religions honor him as the saint for nature and animals.

Like St. Francis, Pope Francis believes that all of nature is connected. He realizes that our Earth and its creatures are suffering, so he wrote a letter to all people in the world to teach that Earth is our common home. In the letter, he described what is wrong and how everyone, including children, can take better care of the Earth and its creatures.

In his letter, the pope said that St. Francis called our planet, "our sister, Mother Earth," but humans have damaged her, forgetting that we are connected to her.

The pope pointed out that even though people were created from Mother Earth, they have forgotten and continue to mistreat her.

His letter explains that because things on Earth are changing so fast, nature can't keep up with them. Air pollution is warming our planet, and there is more and more water and land pollution because of what people are doing.

Today, we use lots of disposable items, mostly made of plastic, and then throw them away. All of this garbage is making it hard for wildlife to find good places to live. It is hurting Mother Earth's natural habitats.

Each year, thousands of plants and animal *species* go *extinct*, mostly because of the harmful things people do. The pope feels sad that children in the future may never see some of the wonders of nature, such as beautiful forests or many of God's creatures, like elephants, lions, tigers, polar bears, and even certain insects.

The pope agrees with scientists who say that the biggest threat to nature comes from humans taking land that wildlife live on in order to build new highways, houses, and shopping malls. People are also taking more and more land for planting crops, which leaves wildlife with less and less room to roam.

The pope also says we must be concerned with the extinction of worms, insects, reptiles, and even tiny creatures we can only see with a microscope. Even though we can't see some of these animals, they are important for maintaining balance in nature.

The letter explains that all of the bad things being done to Mother Earth have caused her to cry out to us, asking us to start doing things differently. It says that God calls us to do His work here on Earth, so that our planet might be what He created it to be: a planet of peace, beauty, and fullness for all of God's creatures—people and animals.

Pope Francis asks that all people throughout the world join with science and work together to fix what is happening to Mother Earth and her creatures. And the good news is that people are beginning to listen. Many people are now working to fix the problems!

Pope Francis's letter frames a very bright future by saying that people need to be thankful for all of the creatures they share the Earth with and join together in recognizing the beauty of creation. He says that if all people work together they can help "make all things new," just as God promised.

Note to Parents to Understand the Papal Encyclical *Laudato Si'*

There has been much attention given to the encyclical, *Laudato Si'*, mostly concerning its treatment of climate change. Historically, encyclicals have been written with a Roman Catholic audience in mind, but this one is unique in that it is addressed to every person on the planet.

Not only is the encyclical unique because it focuses on human abuse of the environment, especially concerning human inducement of global warming, but it also gives a remarkable amount of attention to mankind's treatment of animals. For instance, the word "animal" or "animals" (mentioned thirteen times), appears nearly as much as the word "climate" (mentioned fifteen times). This observation doesn't tell the whole story, however.

The word "creature" is also mentioned nearly eighty times. While human beings, too, are acknowledged as creatures in the encyclical, the use of this term to include both human and non-human creatures is very likely purposeful. Integral ecology, or the interrelatedness between the earth and its inhabitants, is a major theme of the encyclical. It is also in the spirit of St. Francis's (patron saint of ecology and Pope Francis's namesake) respect for the universal kinship of all beings.

The entire document illustrates how certain systems and mindsets threaten all creation, whether or not animals are referenced by name specifically.

Adapted from Akisha Townsend Eaton, ***Why Animal Protection Advocates Are Lauding Pope Francis's Encyclical, 'Laudato Si'***. Worldanimal.net

Family Discussion Questions and Actions

1. What did Pio and Pia say that children can do to help nature heal?
2. What might you and your family begin to do today?
3. From what you discuss as a family, make a list of things you can do at home and at school:
   At Home:

   At School:

4. Go to the blog at *www.mattbergles.com* to post your own ideas about what children and their families can do to help endangered animals and read what others have posted.
5. Search the internet for other actions you can take and list them here.

ACKNOWLEDGMENTS

I acknowledge and thank all people who work passionately to conserve the earth's wildlife and the habitat it needs to survive.
I pray St. Francis is helping us.

ABOUT THE AUTHOR

Matt Bergles is a Colorado native who grew up in Pueblo, where he worked in the steel mill before graduating from South High School and CSU – Pueblo with a degree in social science. He earned a master's degree in U.S. history and a Ph.D. in public affairs at the University of Colorado at Denver, as well as a certificate in alternative dispute resolution at the University of Denver. His doctoral studies focused on the effects of agriculture and suburban/exurban sprawl on wildlife. His dissertation researched how government programs can influence farmers and ranchers to proactively conserve endangered species on their land.

Matt currently teaches at a K-8 Christian/Catholic school in Denver, where he has witnessed firsthand young, inner-city kids' curiosity and their love of animals and nature. He has observed that every time the children are asked what community service project or charitable cause they'd like to be involved with, most of the K-3 students pick something related to pets or wildlife. This natural love and curiosity for animals and nature, as well as Matt's longtime intrigue with Francis of Assisi and the groundbreaking release of the papal encyclical *Laudato Si'* is what led Matt to write *Francis Frames the Future* for young children.

Matt is an independent researcher and advocate for wildlife conservation, especially conservation of prairie habitat. He lives in Denver with his two children, Luke and Mary, dog Samuel, and cat Mojo. Please visit www.mattbergles.com for more information.

ABOUT THE ARTIST

Rob Peters is a freelance illustrator, cartoonist and designer. He has a degree in Visual Communications from Judson College in Elgin, Illinois, and previously worked as a cover artist and designer in the yearbook industry for over five years. As a freelance artist, he has designed logos, book covers, and illustrated over thirty children's picture books, including *Larry Saves the Prairie* and *Go Cars Go!* Rob currently lives in Topeka, Kansas, with his wife and children.

To learn more about Rob, please visit www.rob-peters.com.

ABOUT THE PRESS

Merry Dissonance Press is a book producer/indie publisher of works of transformation, inspiration, exploration, and illumination. MDP takes a holistic approach to bringing books into the world that make a little noise and create dissonance within the whole in order that ALL can be resolved to produce beautiful harmonies.

Merry Dissonance Press works with its authors every step of the way to craft the finest books and help promote them. Dedicated to publishing award-winning books, we strive to support talented writers and assist them to discover, claim, and refine their own distinct voice. Merry Dissonance Press is the place where collaboration and facilitation of our shared human experiences join together to make a difference in our world.

For more information, visit merrydissonancepress.com.